The Planner Belongs To:

____

Helpful Tips

- Do not use personal information
- Do not use real words
- Create longer passwords
- Modify your easy to remember passwords
- Change your password at least every 3 -6 months
- Use different passwords on different accounts
- Keep your passwords protected
- Don't share your passwords

Passwords

NAME:

DATE:

Keep protected & update every 3 - 6 months.

Website	
Username	
Password	
Email	
Notes	

Website	
Username	
Password	
Email	
Notes	

Website	
Username	
Password	
Email	
Notes	

Website	
Username	
Password	
Email	
Notes	

Website	
Username	
Password	
Email	
Notes	

Website	
Username	
Password	
Email	
Notes	

Website	
Username	
Password	
Email	
Notes	

Website	
Username	
Password	
Email	
Notes	

Passwords

NAME:
DATE:

Keep protected & update every 3 - 6 months.

Website	
Username	
Password	
Email	
Notes	

Website	
Username	
Password	
Email	
Notes	

Website	
Username	
Password	
Email	
Notes	

Website	
Username	
Password	
Email	
Notes	

Website	
Username	
Password	
Email	
Notes	

Website	
Username	
Password	
Email	
Notes	

Website	
Username	
Password	
Email	
Notes	

Website	
Username	
Password	
Email	
Notes	

Passwords

NAME:
DATE:

Keep protected & update every 3 - 6 months.

Website	
Username	
Password	
Email	
Notes	

Website	
Username	
Password	
Email	
Notes	

Website	
Username	
Password	
Email	
Notes	

Website	
Username	
Password	
Email	
Notes	

Website	
Username	
Password	
Email	
Notes	

Website	
Username	
Password	
Email	
Notes	

Website	
Username	
Password	
Email	
Notes	

Website	
Username	
Password	
Email	
Notes	

Passwords

NAME:
DATE:

Keep protected & update every 3 - 6 months.

Website	
Username	
Password	
Email	
Notes	

Website	
Username	
Password	
Email	
Notes	

Website	
Username	
Password	
Email	
Notes	

Website	
Username	
Password	
Email	
Notes	

Website	
Username	
Password	
Email	
Notes	

Website	
Username	
Password	
Email	
Notes	

Website	
Username	
Password	
Email	
Notes	

Website	
Username	
Password	
Email	
Notes	

Passwords

NAME:
DATE:

Keep protected & update every 3 - 6 months.

Website	
Username	
Password	
Email	
Notes	

Website	
Username	
Password	
Email	
Notes	

Website	
Username	
Password	
Email	
Notes	

Website	
Username	
Password	
Email	
Notes	

Website	
Username	
Password	
Email	
Notes	

Website	
Username	
Password	
Email	
Notes	

Website	
Username	
Password	
Email	
Notes	

Website	
Username	
Password	
Email	
Notes	

Passwords

NAME:
DATE:

Keep protected & update every 3 - 6 months.

Website	
Username	
Password	
Email	
Notes	

Website	
Username	
Password	
Email	
Notes	

Website	
Username	
Password	
Email	
Notes	

Website	
Username	
Password	
Email	
Notes	

Website	
Username	
Password	
Email	
Notes	

Website	
Username	
Password	
Email	
Notes	

Website	
Username	
Password	
Email	
Notes	

Website	
Username	
Password	
Email	
Notes	

Passwords

NAME:
DATE:

Keep protected & update every 3 - 6 months.

Website	
Username	
Password	
Email	
Notes	

Website	
Username	
Password	
Email	
Notes	

Website	
Username	
Password	
Email	
Notes	

Website	
Username	
Password	
Email	
Notes	

Website	
Username	
Password	
Email	
Notes	

Website	
Username	
Password	
Email	
Notes	

Website	
Username	
Password	
Email	
Notes	

Website	
Username	
Password	
Email	
Notes	

Passwords

NAME:
DATE:

Keep protected & update every 3 - 6 months.

Website	
Username	
Password	
Email	
Notes	

Website	
Username	
Password	
Email	
Notes	

Website	
Username	
Password	
Email	
Notes	

Website	
Username	
Password	
Email	
Notes	

Website	
Username	
Password	
Email	
Notes	

Website	
Username	
Password	
Email	
Notes	

Website	
Username	
Password	
Email	
Notes	

Website	
Username	
Password	
Email	
Notes	

Passwords

NAME:
DATE:

Keep protected & update every 3 - 6 months.

Website	
Username	
Password	
Email	
Notes	

Website	
Username	
Password	
Email	
Notes	

Website	
Username	
Password	
Email	
Notes	

Website	
Username	
Password	
Email	
Notes	

Website	
Username	
Password	
Email	
Notes	

Website	
Username	
Password	
Email	
Notes	

Website	
Username	
Password	
Email	
Notes	

Website	
Username	
Password	
Email	
Notes	

Passwords

NAME:
DATE:

Keep protected & update every 3 - 6 months.

Website	
Username	
Password	
Email	
Notes	

Website	
Username	
Password	
Email	
Notes	

Website	
Username	
Password	
Email	
Notes	

Website	
Username	
Password	
Email	
Notes	

Website	
Username	
Password	
Email	
Notes	

Website	
Username	
Password	
Email	
Notes	

Website	
Username	
Password	
Email	
Notes	

Website	
Username	
Password	
Email	
Notes	

Passwords

NAME:
DATE:

Keep protected & update every 3 - 6 months.

Website	
Username	
Password	
Email	
Notes	

Website	
Username	
Password	
Email	
Notes	

Website	
Username	
Password	
Email	
Notes	

Website	
Username	
Password	
Email	
Notes	

Website	
Username	
Password	
Email	
Notes	

Website	
Username	
Password	
Email	
Notes	

Website	
Username	
Password	
Email	
Notes	

Website	
Username	
Password	
Email	
Notes	

Passwords

NAME:
DATE:

Keep protected & update every 3 - 6 months.

Website	
Username	
Password	
Email	
Notes	

Website	
Username	
Password	
Email	
Notes	

Website	
Username	
Password	
Email	
Notes	

Website	
Username	
Password	
Email	
Notes	

Website	
Username	
Password	
Email	
Notes	

Website	
Username	
Password	
Email	
Notes	

Website	
Username	
Password	
Email	
Notes	

Website	
Username	
Password	
Email	
Notes	

Passwords

NAME:
DATE:

Keep protected & update every 3 - 6 months.

Website	
Username	
Password	
Email	
Notes	

Website	
Username	
Password	
Email	
Notes	

Website	
Username	
Password	
Email	
Notes	

Website	
Username	
Password	
Email	
Notes	

Website	
Username	
Password	
Email	
Notes	

Website	
Username	
Password	
Email	
Notes	

Website	
Username	
Password	
Email	
Notes	

Website	
Username	
Password	
Email	
Notes	

Passwords

NAME:
DATE:

Keep protected & update every 3 - 6 months.

Website	
Username	
Password	
Email	
Notes	

Website	
Username	
Password	
Email	
Notes	

Website	
Username	
Password	
Email	
Notes	

Website	
Username	
Password	
Email	
Notes	

Website	
Username	
Password	
Email	
Notes	

Website	
Username	
Password	
Email	
Notes	

Website	
Username	
Password	
Email	
Notes	

Website	
Username	
Password	
Email	
Notes	

Passwords

NAME:
DATE:

Keep protected & update every 3 - 6 months.

Website	
Username	
Password	
Email	
Notes	

Website	
Username	
Password	
Email	
Notes	

Website	
Username	
Password	
Email	
Notes	

Website	
Username	
Password	
Email	
Notes	

Website	
Username	
Password	
Email	
Notes	

Website	
Username	
Password	
Email	
Notes	

Website	
Username	
Password	
Email	
Notes	

Website	
Username	
Password	
Email	
Notes	

Passwords

NAME:
DATE:

Keep protected & update every 3 - 6 months.

Website	
Username	
Password	
Email	
Notes	

Website	
Username	
Password	
Email	
Notes	

Website	
Username	
Password	
Email	
Notes	

Website	
Username	
Password	
Email	
Notes	

Website	
Username	
Password	
Email	
Notes	

Website	
Username	
Password	
Email	
Notes	

Website	
Username	
Password	
Email	
Notes	

Website	
Username	
Password	
Email	
Notes	

Passwords

NAME:
DATE:

Keep protected & update every 3 - 6 months.

Website	
Username	
Password	
Email	
Notes	

Website	
Username	
Password	
Email	
Notes	

Website	
Username	
Password	
Email	
Notes	

Website	
Username	
Password	
Email	
Notes	

Website	
Username	
Password	
Email	
Notes	

Website	
Username	
Password	
Email	
Notes	

Website	
Username	
Password	
Email	
Notes	

Website	
Username	
Password	
Email	
Notes	

Passwords

NAME:
DATE:

Keep protected & update every 3 - 6 months.

Website	
Username	
Password	
Email	
Notes	

Website	
Username	
Password	
Email	
Notes	

Website	
Username	
Password	
Email	
Notes	

Website	
Username	
Password	
Email	
Notes	

Website	
Username	
Password	
Email	
Notes	

Website	
Username	
Password	
Email	
Notes	

Website	
Username	
Password	
Email	
Notes	

Website	
Username	
Password	
Email	
Notes	

Passwords

NAME:
DATE:

Keep protected & update every 3 - 6 months.

Website	
Username	
Password	
Email	
Notes	

Website	
Username	
Password	
Email	
Notes	

Website	
Username	
Password	
Email	
Notes	

Website	
Username	
Password	
Email	
Notes	

Website	
Username	
Password	
Email	
Notes	

Website	
Username	
Password	
Email	
Notes	

Website	
Username	
Password	
Email	
Notes	

Website	
Username	
Password	
Email	
Notes	

Passwords

NAME:
DATE:

Keep protected & update every 3 - 6 months.

Website	
Username	
Password	
Email	
Notes	

Website	
Username	
Password	
Email	
Notes	

Website	
Username	
Password	
Email	
Notes	

Website	
Username	
Password	
Email	
Notes	

Website	
Username	
Password	
Email	
Notes	

Website	
Username	
Password	
Email	
Notes	

Website	
Username	
Password	
Email	
Notes	

Website	
Username	
Password	
Email	
Notes	

Passwords

NAME:
DATE:

Keep protected & update every 3 - 6 months.

Website	
Username	
Password	
Email	
Notes	

Website	
Username	
Password	
Email	
Notes	

Website	
Username	
Password	
Email	
Notes	

Website	
Username	
Password	
Email	
Notes	

Website	
Username	
Password	
Email	
Notes	

Website	
Username	
Password	
Email	
Notes	

Website	
Username	
Password	
Email	
Notes	

Website	
Username	
Password	
Email	
Notes	

Passwords

NAME:
DATE:

Keep protected & update every 3 - 6 months.

Website	
Username	
Password	
Email	
Notes	

Website	
Username	
Password	
Email	
Notes	

Website	
Username	
Password	
Email	
Notes	

Website	
Username	
Password	
Email	
Notes	

Website	
Username	
Password	
Email	
Notes	

Website	
Username	
Password	
Email	
Notes	

Website	
Username	
Password	
Email	
Notes	

Website	
Username	
Password	
Email	
Notes	

Passwords

NAME:
DATE:

Keep protected & update every 3 - 6 months.

Website	
Username	
Password	
Email	
Notes	

Website	
Username	
Password	
Email	
Notes	

Website	
Username	
Password	
Email	
Notes	

Website	
Username	
Password	
Email	
Notes	

Website	
Username	
Password	
Email	
Notes	

Website	
Username	
Password	
Email	
Notes	

Website	
Username	
Password	
Email	
Notes	

Website	
Username	
Password	
Email	
Notes	

Passwords

NAME:
DATE:

Keep protected & update every 3 - 6 months.

Website	
Username	
Password	
Email	
Notes	

Website	
Username	
Password	
Email	
Notes	

Website	
Username	
Password	
Email	
Notes	

Website	
Username	
Password	
Email	
Notes	

Website	
Username	
Password	
Email	
Notes	

Website	
Username	
Password	
Email	
Notes	

Website	
Username	
Password	
Email	
Notes	

Website	
Username	
Password	
Email	
Notes	

Passwords

NAME:
DATE:

Keep protected & update every 3 - 6 months.

Website	
Username	
Password	
Email	
Notes	

Website	
Username	
Password	
Email	
Notes	

Website	
Username	
Password	
Email	
Notes	

Website	
Username	
Password	
Email	
Notes	

Website	
Username	
Password	
Email	
Notes	

Website	
Username	
Password	
Email	
Notes	

Website	
Username	
Password	
Email	
Notes	

Website	
Username	
Password	
Email	
Notes	

Passwords

NAME:
DATE:

Keep protected & update every 3 - 6 months.

Website	
Username	
Password	
Email	
Notes	

Website	
Username	
Password	
Email	
Notes	

Website	
Username	
Password	
Email	
Notes	

Website	
Username	
Password	
Email	
Notes	

Website	
Username	
Password	
Email	
Notes	

Website	
Username	
Password	
Email	
Notes	

Website	
Username	
Password	
Email	
Notes	

Website	
Username	
Password	
Email	
Notes	

Passwords

NAME:
DATE:

Keep protected & update every 3 - 6 months.

Website	
Username	
Password	
Email	
Notes	

Website	
Username	
Password	
Email	
Notes	

Website	
Username	
Password	
Email	
Notes	

Website	
Username	
Password	
Email	
Notes	

Website	
Username	
Password	
Email	
Notes	

Website	
Username	
Password	
Email	
Notes	

Website	
Username	
Password	
Email	
Notes	

Website	
Username	
Password	
Email	
Notes	

Passwords

NAME:
DATE:

Keep protected & update every 3 - 6 months.

Website	
Username	
Password	
Email	
Notes	

Website	
Username	
Password	
Email	
Notes	

Website	
Username	
Password	
Email	
Notes	

Website	
Username	
Password	
Email	
Notes	

Website	
Username	
Password	
Email	
Notes	

Website	
Username	
Password	
Email	
Notes	

Website	
Username	
Password	
Email	
Notes	

Website	
Username	
Password	
Email	
Notes	

Passwords

NAME:
DATE:

Keep protected & update every 3 - 6 months.

Website	
Username	
Password	
Email	
Notes	

Website	
Username	
Password	
Email	
Notes	

Website	
Username	
Password	
Email	
Notes	

Website	
Username	
Password	
Email	
Notes	

Website	
Username	
Password	
Email	
Notes	

Website	
Username	
Password	
Email	
Notes	

Website	
Username	
Password	
Email	
Notes	

Website	
Username	
Password	
Email	
Notes	

Passwords

NAME:
DATE:

Keep protected & update every 3 - 6 months.

Website	
Username	
Password	
Email	
Notes	

Website	
Username	
Password	
Email	
Notes	

Website	
Username	
Password	
Email	
Notes	

Website	
Username	
Password	
Email	
Notes	

Website	
Username	
Password	
Email	
Notes	

Website	
Username	
Password	
Email	
Notes	

Website	
Username	
Password	
Email	
Notes	

Website	
Username	
Password	
Email	
Notes	

Passwords

NAME:
DATE:

Keep protected & update every 3 - 6 months.

Website	
Username	
Password	
Email	
Notes	

Website	
Username	
Password	
Email	
Notes	

Website	
Username	
Password	
Email	
Notes	

Website	
Username	
Password	
Email	
Notes	

Website	
Username	
Password	
Email	
Notes	

Website	
Username	
Password	
Email	
Notes	

Website	
Username	
Password	
Email	
Notes	

Website	
Username	
Password	
Email	
Notes	

Passwords

NAME:
DATE:

Keep protected & update every 3 - 6 months.

Website	
Username	
Password	
Email	
Notes	

Website	
Username	
Password	
Email	
Notes	

Website	
Username	
Password	
Email	
Notes	

Website	
Username	
Password	
Email	
Notes	

Website	
Username	
Password	
Email	
Notes	

Website	
Username	
Password	
Email	
Notes	

Website	
Username	
Password	
Email	
Notes	

Website	
Username	
Password	
Email	
Notes	

Passwords

NAME:
DATE:

Keep protected & update every 3 - 6 months.

Website	
Username	
Password	
Email	
Notes	

Website	
Username	
Password	
Email	
Notes	

Website	
Username	
Password	
Email	
Notes	

Website	
Username	
Password	
Email	
Notes	

Website	
Username	
Password	
Email	
Notes	

Website	
Username	
Password	
Email	
Notes	

Website	
Username	
Password	
Email	
Notes	

Website	
Username	
Password	
Email	
Notes	

Passwords

NAME:
DATE:

Keep protected & update every 3 - 6 months.

Website	
Username	
Password	
Email	
Notes	

Website	
Username	
Password	
Email	
Notes	

Website	
Username	
Password	
Email	
Notes	

Website	
Username	
Password	
Email	
Notes	

Website	
Username	
Password	
Email	
Notes	

Website	
Username	
Password	
Email	
Notes	

Website	
Username	
Password	
Email	
Notes	

Website	
Username	
Password	
Email	
Notes	

Passwords

NAME:
DATE:

Keep protected & update every 3 - 6 months.

Website	
Username	
Password	
Email	
Notes	

Website	
Username	
Password	
Email	
Notes	

Website	
Username	
Password	
Email	
Notes	

Website	
Username	
Password	
Email	
Notes	

Website	
Username	
Password	
Email	
Notes	

Website	
Username	
Password	
Email	
Notes	

Website	
Username	
Password	
Email	
Notes	

Website	
Username	
Password	
Email	
Notes	

Passwords

NAME:
DATE:

Keep protected & update every 3 - 6 months.

Website	
Username	
Password	
Email	
Notes	

Website	
Username	
Password	
Email	
Notes	

Website	
Username	
Password	
Email	
Notes	

Website	
Username	
Password	
Email	
Notes	

Website	
Username	
Password	
Email	
Notes	

Website	
Username	
Password	
Email	
Notes	

Website	
Username	
Password	
Email	
Notes	

Website	
Username	
Password	
Email	
Notes	

Passwords

NAME:
DATE:

Keep protected & update every 3 - 6 months.

Website	
Username	
Password	
Email	
Notes	

Website	
Username	
Password	
Email	
Notes	

Website	
Username	
Password	
Email	
Notes	

Website	
Username	
Password	
Email	
Notes	

Website	
Username	
Password	
Email	
Notes	

Website	
Username	
Password	
Email	
Notes	

Website	
Username	
Password	
Email	
Notes	

Website	
Username	
Password	
Email	
Notes	

Passwords

NAME:
DATE:

Keep protected & update every 3 - 6 months.

Website	
Username	
Password	
Email	
Notes	

Website	
Username	
Password	
Email	
Notes	

Website	
Username	
Password	
Email	
Notes	

Website	
Username	
Password	
Email	
Notes	

Website	
Username	
Password	
Email	
Notes	

Website	
Username	
Password	
Email	
Notes	

Website	
Username	
Password	
Email	
Notes	

Website	
Username	
Password	
Email	
Notes	

Passwords

NAME:
DATE:

Keep protected & update every 3 - 6 months.

Website	
Username	
Password	
Email	
Notes	

Website	
Username	
Password	
Email	
Notes	

Website	
Username	
Password	
Email	
Notes	

Website	
Username	
Password	
Email	
Notes	

Website	
Username	
Password	
Email	
Notes	

Website	
Username	
Password	
Email	
Notes	

Website	
Username	
Password	
Email	
Notes	

Website	
Username	
Password	
Email	
Notes	

Passwords

NAME:
DATE:

Keep protected & update every 3 - 6 months.

Website	
Username	
Password	
Email	
Notes	

Website	
Username	
Password	
Email	
Notes	

Website	
Username	
Password	
Email	
Notes	

Website	
Username	
Password	
Email	
Notes	

Website	
Username	
Password	
Email	
Notes	

Website	
Username	
Password	
Email	
Notes	

Website	
Username	
Password	
Email	
Notes	

Website	
Username	
Password	
Email	
Notes	

Passwords

NAME:
DATE:

Keep protected & update every 3 - 6 months.

Website	
Username	
Password	
Email	
Notes	

Website	
Username	
Password	
Email	
Notes	

Website	
Username	
Password	
Email	
Notes	

Website	
Username	
Password	
Email	
Notes	

Website	
Username	
Password	
Email	
Notes	

Website	
Username	
Password	
Email	
Notes	

Website	
Username	
Password	
Email	
Notes	

Website	
Username	
Password	
Email	
Notes	

Passwords

NAME:
DATE:

Keep protected & update every 3 - 6 months.

Website	
Username	
Password	
Email	
Notes	

Website	
Username	
Password	
Email	
Notes	

Website	
Username	
Password	
Email	
Notes	

Website	
Username	
Password	
Email	
Notes	

Website	
Username	
Password	
Email	
Notes	

Website	
Username	
Password	
Email	
Notes	

Website	
Username	
Password	
Email	
Notes	

Website	
Username	
Password	
Email	
Notes	

Passwords

NAME:
DATE:

Keep protected & update every 3 - 6 months.

Website	
Username	
Password	
Email	
Notes	

Website	
Username	
Password	
Email	
Notes	

Website	
Username	
Password	
Email	
Notes	

Website	
Username	
Password	
Email	
Notes	

Website	
Username	
Password	
Email	
Notes	

Website	
Username	
Password	
Email	
Notes	

Website	
Username	
Password	
Email	
Notes	

Website	
Username	
Password	
Email	
Notes	

Passwords

NAME:
DATE:

Keep protected & update every 3 - 6 months.

Website	
Username	
Password	
Email	
Notes	

Website	
Username	
Password	
Email	
Notes	

Website	
Username	
Password	
Email	
Notes	

Website	
Username	
Password	
Email	
Notes	

Website	
Username	
Password	
Email	
Notes	

Website	
Username	
Password	
Email	
Notes	

Website	
Username	
Password	
Email	
Notes	

Website	
Username	
Password	
Email	
Notes	

Passwords

NAME:
DATE:

Keep protected & update every 3 - 6 months.

Website	
Username	
Password	
Email	
Notes	

Website	
Username	
Password	
Email	
Notes	

Website	
Username	
Password	
Email	
Notes	

Website	
Username	
Password	
Email	
Notes	

Website	
Username	
Password	
Email	
Notes	

Website	
Username	
Password	
Email	
Notes	

Website	
Username	
Password	
Email	
Notes	

Website	
Username	
Password	
Email	
Notes	

Passwords

NAME:
DATE:

Keep protected & update every 3 - 6 months.

Website	
Username	
Password	
Email	
Notes	

Website	
Username	
Password	
Email	
Notes	

Website	
Username	
Password	
Email	
Notes	

Website	
Username	
Password	
Email	
Notes	

Website	
Username	
Password	
Email	
Notes	

Website	
Username	
Password	
Email	
Notes	

Website	
Username	
Password	
Email	
Notes	

Website	
Username	
Password	
Email	
Notes	

Passwords

NAME:
DATE:

Keep protected & update every 3 - 6 months.

Website	
Username	
Password	
Email	
Notes	

Website	
Username	
Password	
Email	
Notes	

Website	
Username	
Password	
Email	
Notes	

Website	
Username	
Password	
Email	
Notes	

Website	
Username	
Password	
Email	
Notes	

Website	
Username	
Password	
Email	
Notes	

Website	
Username	
Password	
Email	
Notes	

Website	
Username	
Password	
Email	
Notes	

Passwords

NAME:
DATE:

Keep protected & update every 3 - 6 months.

Website	
Username	
Password	
Email	
Notes	

Website	
Username	
Password	
Email	
Notes	

Website	
Username	
Password	
Email	
Notes	

Website	
Username	
Password	
Email	
Notes	

Website	
Username	
Password	
Email	
Notes	

Website	
Username	
Password	
Email	
Notes	

Website	
Username	
Password	
Email	
Notes	

Website	
Username	
Password	
Email	
Notes	

Passwords

NAME:

DATE:

Keep protected & update every 3 - 6 months.

Website	
Username	
Password	
Email	
Notes	

Website	
Username	
Password	
Email	
Notes	

Website	
Username	
Password	
Email	
Notes	

Website	
Username	
Password	
Email	
Notes	

Website	
Username	
Password	
Email	
Notes	

Website	
Username	
Password	
Email	
Notes	

Website	
Username	
Password	
Email	
Notes	

Website	
Username	
Password	
Email	
Notes	

Passwords

NAME:
DATE:

Keep protected & update every 3 - 6 months.

Website	
Username	
Password	
Email	
Notes	

Website	
Username	
Password	
Email	
Notes	

Website	
Username	
Password	
Email	
Notes	

Website	
Username	
Password	
Email	
Notes	

Website	
Username	
Password	
Email	
Notes	

Website	
Username	
Password	
Email	
Notes	

Website	
Username	
Password	
Email	
Notes	

Website	
Username	
Password	
Email	
Notes	

Passwords

NAME:
DATE:

Keep protected & update every 3 - 6 months.

Website	
Username	
Password	
Email	
Notes	

Website	
Username	
Password	
Email	
Notes	

Website	
Username	
Password	
Email	
Notes	

Website	
Username	
Password	
Email	
Notes	

Website	
Username	
Password	
Email	
Notes	

Website	
Username	
Password	
Email	
Notes	

Website	
Username	
Password	
Email	
Notes	

Website	
Username	
Password	
Email	
Notes	

Passwords

NAME:
DATE:

Keep protected & update every 3 - 6 months.

Website	
Username	
Password	
Email	
Notes	

Website	
Username	
Password	
Email	
Notes	

Website	
Username	
Password	
Email	
Notes	

Website	
Username	
Password	
Email	
Notes	

Website	
Username	
Password	
Email	
Notes	

Website	
Username	
Password	
Email	
Notes	

Website	
Username	
Password	
Email	
Notes	

Website	
Username	
Password	
Email	
Notes	

Passwords

NAME:
DATE:

Keep protected & update every 3 - 6 months.

Website	
Username	
Password	
Email	
Notes	

Website	
Username	
Password	
Email	
Notes	

Website	
Username	
Password	
Email	
Notes	

Website	
Username	
Password	
Email	
Notes	

Website	
Username	
Password	
Email	
Notes	

Website	
Username	
Password	
Email	
Notes	

Website	
Username	
Password	
Email	
Notes	

Website	
Username	
Password	
Email	
Notes	

Passwords

NAME:
DATE:

Keep protected & update every 3 - 6 months.

Website	
Username	
Password	
Email	
Notes	

Website	
Username	
Password	
Email	
Notes	

Website	
Username	
Password	
Email	
Notes	

Website	
Username	
Password	
Email	
Notes	

Website	
Username	
Password	
Email	
Notes	

Website	
Username	
Password	
Email	
Notes	

Website	
Username	
Password	
Email	
Notes	

Website	
Username	
Password	
Email	
Notes	

Passwords

NAME:
DATE:

Keep protected & update every 3 - 6 months.

Website	
Username	
Password	
Email	
Notes	

Website	
Username	
Password	
Email	
Notes	

Website	
Username	
Password	
Email	
Notes	

Website	
Username	
Password	
Email	
Notes	

Website	
Username	
Password	
Email	
Notes	

Website	
Username	
Password	
Email	
Notes	

Website	
Username	
Password	
Email	
Notes	

Website	
Username	
Password	
Email	
Notes	

Passwords

NAME:
DATE:

Keep protected & update every 3 - 6 months.

Website	
Username	
Password	
Email	
Notes	

Website	
Username	
Password	
Email	
Notes	

Website	
Username	
Password	
Email	
Notes	

Website	
Username	
Password	
Email	
Notes	

Website	
Username	
Password	
Email	
Notes	

Website	
Username	
Password	
Email	
Notes	

Website	
Username	
Password	
Email	
Notes	

Website	
Username	
Password	
Email	
Notes	

Passwords

NAME:
DATE:

Keep protected & update every 3 - 6 months.

Website	
Username	
Password	
Email	
Notes	

Website	
Username	
Password	
Email	
Notes	

Website	
Username	
Password	
Email	
Notes	

Website	
Username	
Password	
Email	
Notes	

Website	
Username	
Password	
Email	
Notes	

Website	
Username	
Password	
Email	
Notes	

Website	
Username	
Password	
Email	
Notes	

Website	
Username	
Password	
Email	
Notes	

Passwords

NAME:

DATE:

Keep protected & update every 3 - 6 months.

Website	
Username	
Password	
Email	
Notes	

Website	
Username	
Password	
Email	
Notes	

Website	
Username	
Password	
Email	
Notes	

Website	
Username	
Password	
Email	
Notes	

Website	
Username	
Password	
Email	
Notes	

Website	
Username	
Password	
Email	
Notes	

Website	
Username	
Password	
Email	
Notes	

Website	
Username	
Password	
Email	
Notes	

Passwords

NAME:
DATE:

Keep protected & update every 3 - 6 months.

Website	
Username	
Password	
Email	
Notes	

Website	
Username	
Password	
Email	
Notes	

Website	
Username	
Password	
Email	
Notes	

Website	
Username	
Password	
Email	
Notes	

Website	
Username	
Password	
Email	
Notes	

Website	
Username	
Password	
Email	
Notes	

Website	
Username	
Password	
Email	
Notes	

Website	
Username	
Password	
Email	
Notes	

Passwords

NAME:
DATE:

Keep protected & update every 3 - 6 months.

Website	
Username	
Password	
Email	
Notes	

Website	
Username	
Password	
Email	
Notes	

Website	
Username	
Password	
Email	
Notes	

Website	
Username	
Password	
Email	
Notes	

Website	
Username	
Password	
Email	
Notes	

Website	
Username	
Password	
Email	
Notes	

Website	
Username	
Password	
Email	
Notes	

Website	
Username	
Password	
Email	
Notes	

Passwords

NAME:
DATE:

Keep protected & update every 3 - 6 months.

Website	
Username	
Password	
Email	
Notes	

Website	
Username	
Password	
Email	
Notes	

Website	
Username	
Password	
Email	
Notes	

Website	
Username	
Password	
Email	
Notes	

Website	
Username	
Password	
Email	
Notes	

Website	
Username	
Password	
Email	
Notes	

Website	
Username	
Password	
Email	
Notes	

Website	
Username	
Password	
Email	
Notes	

Passwords

NAME:
DATE:

Keep protected & update every 3 - 6 months.

Website	
Username	
Password	
Email	
Notes	

Website	
Username	
Password	
Email	
Notes	

Website	
Username	
Password	
Email	
Notes	

Website	
Username	
Password	
Email	
Notes	

Website	
Username	
Password	
Email	
Notes	

Website	
Username	
Password	
Email	
Notes	

Website	
Username	
Password	
Email	
Notes	

Website	
Username	
Password	
Email	
Notes	

Passwords

NAME:
DATE:

Keep protected & update every 3 - 6 months.

Website	
Username	
Password	
Email	
Notes	

Website	
Username	
Password	
Email	
Notes	

Website	
Username	
Password	
Email	
Notes	

Website	
Username	
Password	
Email	
Notes	

Website	
Username	
Password	
Email	
Notes	

Website	
Username	
Password	
Email	
Notes	

Website	
Username	
Password	
Email	
Notes	

Website	
Username	
Password	
Email	
Notes	

Passwords

NAME:
DATE:

Keep protected & update every 3 - 6 months.

Website	
Username	
Password	
Email	
Notes	

Website	
Username	
Password	
Email	
Notes	

Website	
Username	
Password	
Email	
Notes	

Website	
Username	
Password	
Email	
Notes	

Website	
Username	
Password	
Email	
Notes	

Website	
Username	
Password	
Email	
Notes	

Website	
Username	
Password	
Email	
Notes	

Website	
Username	
Password	
Email	
Notes	

Passwords

NAME:
DATE:

Keep protected & update every 3 - 6 months.

Website	
Username	
Password	
Email	
Notes	

Website	
Username	
Password	
Email	
Notes	

Website	
Username	
Password	
Email	
Notes	

Website	
Username	
Password	
Email	
Notes	

Website	
Username	
Password	
Email	
Notes	

Website	
Username	
Password	
Email	
Notes	

Website	
Username	
Password	
Email	
Notes	

Website	
Username	
Password	
Email	
Notes	

Passwords

NAME:
DATE:

Keep protected & update every 3 - 6 months.

Website	
Username	
Password	
Email	
Notes	

Website	
Username	
Password	
Email	
Notes	

Website	
Username	
Password	
Email	
Notes	

Website	
Username	
Password	
Email	
Notes	

Website	
Username	
Password	
Email	
Notes	

Website	
Username	
Password	
Email	
Notes	

Website	
Username	
Password	
Email	
Notes	

Website	
Username	
Password	
Email	
Notes	

Passwords

NAME:
DATE:

Keep protected & update every 3 - 6 months.

Website	
Username	
Password	
Email	
Notes	

Website	
Username	
Password	
Email	
Notes	

Website	
Username	
Password	
Email	
Notes	

Website	
Username	
Password	
Email	
Notes	

Website	
Username	
Password	
Email	
Notes	

Website	
Username	
Password	
Email	
Notes	

Website	
Username	
Password	
Email	
Notes	

Website	
Username	
Password	
Email	
Notes	

Passwords

NAME:
DATE:

Keep protected & update every 3 - 6 months.

Website	
Username	
Password	
Email	
Notes	

Website	
Username	
Password	
Email	
Notes	

Website	
Username	
Password	
Email	
Notes	

Website	
Username	
Password	
Email	
Notes	

Website	
Username	
Password	
Email	
Notes	

Website	
Username	
Password	
Email	
Notes	

Website	
Username	
Password	
Email	
Notes	

Website	
Username	
Password	
Email	
Notes	

Passwords

NAME:
DATE:

Keep protected & update every 3 - 6 months.

Website	
Username	
Password	
Email	
Notes	

Website	
Username	
Password	
Email	
Notes	

Website	
Username	
Password	
Email	
Notes	

Website	
Username	
Password	
Email	
Notes	

Website	
Username	
Password	
Email	
Notes	

Website	
Username	
Password	
Email	
Notes	

Website	
Username	
Password	
Email	
Notes	

Website	
Username	
Password	
Email	
Notes	

Passwords

NAME:
DATE:

Keep protected & update every 3 - 6 months.

Website	
Username	
Password	
Email	
Notes	

Website	
Username	
Password	
Email	
Notes	

Website	
Username	
Password	
Email	
Notes	

Website	
Username	
Password	
Email	
Notes	

Website	
Username	
Password	
Email	
Notes	

Website	
Username	
Password	
Email	
Notes	

Website	
Username	
Password	
Email	
Notes	

Website	
Username	
Password	
Email	
Notes	

Passwords

NAME:
DATE:

Keep protected & update every 3 - 6 months.

Website	
Username	
Password	
Email	
Notes	

Website	
Username	
Password	
Email	
Notes	

Website	
Username	
Password	
Email	
Notes	

Website	
Username	
Password	
Email	
Notes	

Website	
Username	
Password	
Email	
Notes	

Website	
Username	
Password	
Email	
Notes	

Website	
Username	
Password	
Email	
Notes	

Website	
Username	
Password	
Email	
Notes	

Passwords

NAME:
DATE:

Keep protected & update every 3 - 6 months.

Website	
Username	
Password	
Email	
Notes	

Website	
Username	
Password	
Email	
Notes	

Website	
Username	
Password	
Email	
Notes	

Website	
Username	
Password	
Email	
Notes	

Website	
Username	
Password	
Email	
Notes	

Website	
Username	
Password	
Email	
Notes	

Website	
Username	
Password	
Email	
Notes	

Website	
Username	
Password	
Email	
Notes	

Passwords

NAME:
DATE:

Keep protected & update every 3 - 6 months.

Website	
Username	
Password	
Email	
Notes	

Website	
Username	
Password	
Email	
Notes	

Website	
Username	
Password	
Email	
Notes	

Website	
Username	
Password	
Email	
Notes	

Website	
Username	
Password	
Email	
Notes	

Website	
Username	
Password	
Email	
Notes	

Website	
Username	
Password	
Email	
Notes	

Website	
Username	
Password	
Email	
Notes	

Passwords

NAME:
DATE:

Keep protected & update every 3 - 6 months.

Website	
Username	
Password	
Email	
Notes	

Website	
Username	
Password	
Email	
Notes	

Website	
Username	
Password	
Email	
Notes	

Website	
Username	
Password	
Email	
Notes	

Website	
Username	
Password	
Email	
Notes	

Website	
Username	
Password	
Email	
Notes	

Website	
Username	
Password	
Email	
Notes	

Website	
Username	
Password	
Email	
Notes	

Passwords

NAME:
DATE:

Keep protected & update every 3 - 6 months.

Website	
Username	
Password	
Email	
Notes	

Website	
Username	
Password	
Email	
Notes	

Website	
Username	
Password	
Email	
Notes	

Website	
Username	
Password	
Email	
Notes	

Website	
Username	
Password	
Email	
Notes	

Website	
Username	
Password	
Email	
Notes	

Website	
Username	
Password	
Email	
Notes	

Website	
Username	
Password	
Email	
Notes	

Passwords

NAME:
DATE:

Keep protected & update every 3 - 6 months.

Website	
Username	
Password	
Email	
Notes	

Website	
Username	
Password	
Email	
Notes	

Website	
Username	
Password	
Email	
Notes	

Website	
Username	
Password	
Email	
Notes	

Website	
Username	
Password	
Email	
Notes	

Website	
Username	
Password	
Email	
Notes	

Website	
Username	
Password	
Email	
Notes	

Website	
Username	
Password	
Email	
Notes	

Passwords

NAME:
DATE:

Keep protected & update every 3 - 6 months.

Website	
Username	
Password	
Email	
Notes	

Website	
Username	
Password	
Email	
Notes	

Website	
Username	
Password	
Email	
Notes	

Website	
Username	
Password	
Email	
Notes	

Website	
Username	
Password	
Email	
Notes	

Website	
Username	
Password	
Email	
Notes	

Website	
Username	
Password	
Email	
Notes	

Website	
Username	
Password	
Email	
Notes	

Passwords

NAME:
DATE:

Keep protected & update every 3 - 6 months.

Website	
Username	
Password	
Email	
Notes	

Website	
Username	
Password	
Email	
Notes	

Website	
Username	
Password	
Email	
Notes	

Website	
Username	
Password	
Email	
Notes	

Website	
Username	
Password	
Email	
Notes	

Website	
Username	
Password	
Email	
Notes	

Website	
Username	
Password	
Email	
Notes	

Website	
Username	
Password	
Email	
Notes	

Passwords

NAME:
DATE:

Keep protected & update every 3 - 6 months.

Website	
Username	
Password	
Email	
Notes	

Website	
Username	
Password	
Email	
Notes	

Website	
Username	
Password	
Email	
Notes	

Website	
Username	
Password	
Email	
Notes	

Website	
Username	
Password	
Email	
Notes	

Website	
Username	
Password	
Email	
Notes	

Website	
Username	
Password	
Email	
Notes	

Website	
Username	
Password	
Email	
Notes	

Passwords

NAME:
DATE:

Keep protected & update every 3 - 6 months.

Website	
Username	
Password	
Email	
Notes	

Website	
Username	
Password	
Email	
Notes	

Website	
Username	
Password	
Email	
Notes	

Website	
Username	
Password	
Email	
Notes	

Website	
Username	
Password	
Email	
Notes	

Website	
Username	
Password	
Email	
Notes	

Website	
Username	
Password	
Email	
Notes	

Website	
Username	
Password	
Email	
Notes	

Passwords

NAME:
DATE:

Keep protected & update every 3 - 6 months.

Website	
Username	
Password	
Email	
Notes	

Website	
Username	
Password	
Email	
Notes	

Website	
Username	
Password	
Email	
Notes	

Website	
Username	
Password	
Email	
Notes	

Website	
Username	
Password	
Email	
Notes	

Website	
Username	
Password	
Email	
Notes	

Website	
Username	
Password	
Email	
Notes	

Website	
Username	
Password	
Email	
Notes	

Passwords

NAME:
DATE:

Keep protected & update every 3 - 6 months.

Website	
Username	
Password	
Email	
Notes	

Website	
Username	
Password	
Email	
Notes	

Website	
Username	
Password	
Email	
Notes	

Website	
Username	
Password	
Email	
Notes	

Website	
Username	
Password	
Email	
Notes	

Website	
Username	
Password	
Email	
Notes	

Website	
Username	
Password	
Email	
Notes	

Website	
Username	
Password	
Email	
Notes	

Passwords

NAME:
DATE:

Keep protected & update every 3 - 6 months.

Website	
Username	
Password	
Email	
Notes	

Website	
Username	
Password	
Email	
Notes	

Website	
Username	
Password	
Email	
Notes	

Website	
Username	
Password	
Email	
Notes	

Website	
Username	
Password	
Email	
Notes	

Website	
Username	
Password	
Email	
Notes	

Website	
Username	
Password	
Email	
Notes	

Website	
Username	
Password	
Email	
Notes	

Passwords

NAME:
DATE:

Keep protected & update every 3 - 6 months.

Website	
Username	
Password	
Email	
Notes	

Website	
Username	
Password	
Email	
Notes	

Website	
Username	
Password	
Email	
Notes	

Website	
Username	
Password	
Email	
Notes	

Website	
Username	
Password	
Email	
Notes	

Website	
Username	
Password	
Email	
Notes	

Website	
Username	
Password	
Email	
Notes	

Website	
Username	
Password	
Email	
Notes	

Passwords

NAME:
DATE:

Keep protected & update every 3 - 6 months.

Website	
Username	
Password	
Email	
Notes	

Website	
Username	
Password	
Email	
Notes	

Website	
Username	
Password	
Email	
Notes	

Website	
Username	
Password	
Email	
Notes	

Website	
Username	
Password	
Email	
Notes	

Website	
Username	
Password	
Email	
Notes	

Website	
Username	
Password	
Email	
Notes	

Website	
Username	
Password	
Email	
Notes	

Passwords

NAME:
DATE:

Keep protected & update every 3 - 6 months.

Website	
Username	
Password	
Email	
Notes	

Website	
Username	
Password	
Email	
Notes	

Website	
Username	
Password	
Email	
Notes	

Website	
Username	
Password	
Email	
Notes	

Website	
Username	
Password	
Email	
Notes	

Website	
Username	
Password	
Email	
Notes	

Website	
Username	
Password	
Email	
Notes	

Website	
Username	
Password	
Email	
Notes	

Passwords

NAME:
DATE:

Keep protected & update every 3 - 6 months.

Website	
Username	
Password	
Email	
Notes	

Website	
Username	
Password	
Email	
Notes	

Website	
Username	
Password	
Email	
Notes	

Website	
Username	
Password	
Email	
Notes	

Website	
Username	
Password	
Email	
Notes	

Website	
Username	
Password	
Email	
Notes	

Website	
Username	
Password	
Email	
Notes	

Website	
Username	
Password	
Email	
Notes	

Passwords

NAME:
DATE:

Keep protected & update every 3 - 6 months.

Website	
Username	
Password	
Email	
Notes	

Website	
Username	
Password	
Email	
Notes	

Website	
Username	
Password	
Email	
Notes	

Website	
Username	
Password	
Email	
Notes	

Website	
Username	
Password	
Email	
Notes	

Website	
Username	
Password	
Email	
Notes	

Website	
Username	
Password	
Email	
Notes	

Website	
Username	
Password	
Email	
Notes	

Passwords

NAME:
DATE:

Keep protected & update every 3 - 6 months.

Website	
Username	
Password	
Email	
Notes	

Website	
Username	
Password	
Email	
Notes	

Website	
Username	
Password	
Email	
Notes	

Website	
Username	
Password	
Email	
Notes	

Website	
Username	
Password	
Email	
Notes	

Website	
Username	
Password	
Email	
Notes	

Website	
Username	
Password	
Email	
Notes	

Website	
Username	
Password	
Email	
Notes	

Passwords

NAME:
DATE:

Keep protected & update every 3 - 6 months.

Website	
Username	
Password	
Email	
Notes	

Website	
Username	
Password	
Email	
Notes	

Website	
Username	
Password	
Email	
Notes	

Website	
Username	
Password	
Email	
Notes	

Website	
Username	
Password	
Email	
Notes	

Website	
Username	
Password	
Email	
Notes	

Website	
Username	
Password	
Email	
Notes	

Website	
Username	
Password	
Email	
Notes	

Passwords

NAME:
DATE:

Keep protected & update every 3 - 6 months.

Website	
Username	
Password	
Email	
Notes	

Website	
Username	
Password	
Email	
Notes	

Website	
Username	
Password	
Email	
Notes	

Website	
Username	
Password	
Email	
Notes	

Website	
Username	
Password	
Email	
Notes	

Website	
Username	
Password	
Email	
Notes	

Website	
Username	
Password	
Email	
Notes	

Website	
Username	
Password	
Email	
Notes	

Passwords

NAME:
DATE:

Keep protected & update every 3 - 6 months.

Website	
Username	
Password	
Email	
Notes	

Website	
Username	
Password	
Email	
Notes	

Website	
Username	
Password	
Email	
Notes	

Website	
Username	
Password	
Email	
Notes	

Website	
Username	
Password	
Email	
Notes	

Website	
Username	
Password	
Email	
Notes	

Website	
Username	
Password	
Email	
Notes	

Website	
Username	
Password	
Email	
Notes	

Passwords

NAME:
DATE:

Keep protected & update every 3 - 6 months.

Website	
Username	
Password	
Email	
Notes	

Website	
Username	
Password	
Email	
Notes	

Website	
Username	
Password	
Email	
Notes	

Website	
Username	
Password	
Email	
Notes	

Website	
Username	
Password	
Email	
Notes	

Website	
Username	
Password	
Email	
Notes	

Website	
Username	
Password	
Email	
Notes	

Website	
Username	
Password	
Email	
Notes	

Passwords

NAME:
DATE:

Keep protected & update every 3 - 6 months.

Website	
Username	
Password	
Email	
Notes	

Website	
Username	
Password	
Email	
Notes	

Website	
Username	
Password	
Email	
Notes	

Website	
Username	
Password	
Email	
Notes	

Website	
Username	
Password	
Email	
Notes	

Website	
Username	
Password	
Email	
Notes	

Website	
Username	
Password	
Email	
Notes	

Website	
Username	
Password	
Email	
Notes	

Passwords

NAME:
DATE:

Keep protected & update every 3 - 6 months.

Website	
Username	
Password	
Email	
Notes	

Website	
Username	
Password	
Email	
Notes	

Website	
Username	
Password	
Email	
Notes	

Website	
Username	
Password	
Email	
Notes	

Website	
Username	
Password	
Email	
Notes	

Website	
Username	
Password	
Email	
Notes	

Website	
Username	
Password	
Email	
Notes	

Website	
Username	
Password	
Email	
Notes	

Passwords

NAME:
DATE:

Keep protected & update every 3 - 6 months.

Website	
Username	
Password	
Email	
Notes	

Website	
Username	
Password	
Email	
Notes	

Website	
Username	
Password	
Email	
Notes	

Website	
Username	
Password	
Email	
Notes	

Website	
Username	
Password	
Email	
Notes	

Website	
Username	
Password	
Email	
Notes	

Website	
Username	
Password	
Email	
Notes	

Website	
Username	
Password	
Email	
Notes	

Passwords

NAME:
DATE:

Keep protected & update every 3 - 6 months.

Website	
Username	
Password	
Email	
Notes	

Website	
Username	
Password	
Email	
Notes	

Website	
Username	
Password	
Email	
Notes	

Website	
Username	
Password	
Email	
Notes	

Website	
Username	
Password	
Email	
Notes	

Website	
Username	
Password	
Email	
Notes	

Website	
Username	
Password	
Email	
Notes	

Website	
Username	
Password	
Email	
Notes	

Passwords

NAME:
DATE:

Keep protected & update every 3 - 6 months.

Website	
Username	
Password	
Email	
Notes	

Website	
Username	
Password	
Email	
Notes	

Website	
Username	
Password	
Email	
Notes	

Website	
Username	
Password	
Email	
Notes	

Website	
Username	
Password	
Email	
Notes	

Website	
Username	
Password	
Email	
Notes	

Website	
Username	
Password	
Email	
Notes	

Website	
Username	
Password	
Email	
Notes	

Passwords

NAME:
DATE:

Keep protected & update every 3 - 6 months.

Website	
Username	
Password	
Email	
Notes	

Website	
Username	
Password	
Email	
Notes	

Website	
Username	
Password	
Email	
Notes	

Website	
Username	
Password	
Email	
Notes	

Website	
Username	
Password	
Email	
Notes	

Website	
Username	
Password	
Email	
Notes	

Website	
Username	
Password	
Email	
Notes	

Website	
Username	
Password	
Email	
Notes	

Passwords

NAME:
DATE:

Keep protected & update every 3 - 6 months.

Website	
Username	
Password	
Email	
Notes	

Website	
Username	
Password	
Email	
Notes	

Website	
Username	
Password	
Email	
Notes	

Website	
Username	
Password	
Email	
Notes	

Website	
Username	
Password	
Email	
Notes	

Website	
Username	
Password	
Email	
Notes	

Website	
Username	
Password	
Email	
Notes	

Website	
Username	
Password	
Email	
Notes	

Passwords

NAME:
DATE:

Keep protected & update every 3 - 6 months.

Website	
Username	
Password	
Email	
Notes	

Website	
Username	
Password	
Email	
Notes	

Website	
Username	
Password	
Email	
Notes	

Website	
Username	
Password	
Email	
Notes	

Website	
Username	
Password	
Email	
Notes	

Website	
Username	
Password	
Email	
Notes	

Website	
Username	
Password	
Email	
Notes	

Website	
Username	
Password	
Email	
Notes	

Passwords

NAME:
DATE:

Keep protected & update every 3 - 6 months.

Website	
Username	
Password	
Email	
Notes	

Website	
Username	
Password	
Email	
Notes	

Website	
Username	
Password	
Email	
Notes	

Website	
Username	
Password	
Email	
Notes	

Website	
Username	
Password	
Email	
Notes	

Website	
Username	
Password	
Email	
Notes	

Website	
Username	
Password	
Email	
Notes	

Website	
Username	
Password	
Email	
Notes	

Passwords

NAME:
DATE:

Keep protected & update every 3 - 6 months.

Website	
Username	
Password	
Email	
Notes	

Website	
Username	
Password	
Email	
Notes	

Website	
Username	
Password	
Email	
Notes	

Website	
Username	
Password	
Email	
Notes	

Website	
Username	
Password	
Email	
Notes	

Website	
Username	
Password	
Email	
Notes	

Website	
Username	
Password	
Email	
Notes	

Website	
Username	
Password	
Email	
Notes	

Passwords

NAME:
DATE:

Keep protected & update every 3 - 6 months.

Website	
Username	
Password	
Email	
Notes	

Website	
Username	
Password	
Email	
Notes	

Website	
Username	
Password	
Email	
Notes	

Website	
Username	
Password	
Email	
Notes	

Website	
Username	
Password	
Email	
Notes	

Website	
Username	
Password	
Email	
Notes	

Website	
Username	
Password	
Email	
Notes	

Website	
Username	
Password	
Email	
Notes	

Passwords

NAME:
DATE:

Keep protected & update every 3 - 6 months.

Website	
Username	
Password	
Email	
Notes	

Website	
Username	
Password	
Email	
Notes	

Website	
Username	
Password	
Email	
Notes	

Website	
Username	
Password	
Email	
Notes	

Website	
Username	
Password	
Email	
Notes	

Website	
Username	
Password	
Email	
Notes	

Website	
Username	
Password	
Email	
Notes	

Website	
Username	
Password	
Email	
Notes	

NOTES

NOTES

NOTES

NOTES

NOTES

NOTES

NOTES

NOTES

NOTES

NOTES

Made in the USA
Monee, IL
04 November 2026

67352869R00069